GOD THE CREATOR

TOWARD A MORE ROBUST DOCTRINE OF CREATION

Henry E. Neufeld

Topical Line Drives, Volume 6

Energion Publications

Gonzalez, FL

2020

Unless otherwise noted, translations are by the author.

ISBN10: 1-938434-74-9
ISBN13: 978-1-938434-74-7
eISBN: 978-1-938434-90-7

Energion Publications
P. O. Box 841
Gonzalez, FL 32560

pubs@energion.com

Dedication

To James Kristian McClellan.
Thank you for seeing things differently!

Introduction

While Christians are unified in seeing God as creator, they hold many different views about how God created the world. I frequently encounter people who are surprised to discover that different views exist. This isn't just on one side. I find it equally likely to encounter a young earth creationist who is shocked that there are people who call themselves Christians and who also accept the theory of evolution as I am to find those who accept evolution but can't believe there are people who actually accept a young earth (less than 10,000 years old) and a literal, seven day creation week.

In this small book I'm going to start by outlining the major views and key arguments. Much of this material will duplicate material found in my FastTracts book *Christian Views on Origins*.[1]

Differences in Sources

One of the reasons debates are so contentious is that this issue has important implications. It is understandable that people are concerned with God's relationship to the universe and to them in particular. That is, after all, what the Christian message is about.

But what makes arguments even more contentious is that we disagree about what our sources of knowledge are, and how those sources are to be used.

This begins with basic differences in what we expect from our sources. Theologians distinguish general revelation, which is what is revealed in nature, in creation in general, and special revelation, which is what God has provided through his word in the form of the scriptures. There is a difference in the weight that different people give to these two sources. How much information does scripture provide about the physical world, and how are we to relate it to what we derive from science, i.e., from observation? This

---

1 Henry Neufeld, ed, *Christian Views on Origins*. Gonzalez, Florida: Energion Publications, 2020.

is very contentious. Both theistic evolutionists and young earth creationists will claim that the Bible and science do not conflict, but they mean very different things by those claims.

It is not simply a matter of the weight given to each, but the approach one takes to understanding each, and especially to scripture. Do you see scripture as a collection of various types of literature, or is it all similar? If you can read something as a poetic or even fictional account, for example, it can be of value for something, but generally not as an expression of scientific fact.

We will look more closely both at what we can expect to discover from various sources, how we can expect to discover it, and what weight to give the result.

DIFFERENCES IN PRIORITIES

Besides our differences regarding the information, we also differ on the implications of that information. There are those who maintain that a belief in evolutionary theory really makes no difference theologically. This is not the case. It is both rooted in and results in a difference in our view of God, and especially of revelation. This can impact the way we read both general and special revelation.

Those who maintain a young earth position generally believe this is the only way to uphold the veracity and authority of scripture. Thus creation becomes tied to the authority of God's word in scripture and to the accuracy of revelation. This leads back to what kind of literature one is reading. We would not criticize a historical novel because a character in the story did not exist historically. In such a novel, some of the characters are drawn from history and some are not.

To the believer in a young age for the universe, certain passages are narrative history, so they see anything that challenges that narrative history as contrary to scripture. A theistic evolutionist would likely claim to be honoring that scripture, but as a different literary genre, such as liturgy, myth[2], or parable.

On the other hand there are those who work primarily from a theological view of salvation history. God the creator is also God

---

2 In a literary sense.

2

the redeemer, and the issue is much more one of tying the first and second Adams together. This raises the additional question of whether a historical Adam and Eve are required in order to believe in the story of redemption.

The choices are not binary. There are a range of key issues (such as the existence of a historical Adam, or a precisely seven day creation week) which a particular believer holds as essential beliefs. When one challenges these essential beliefs, the person holding them feels the foundations of their faith shake. If X is no longer true, then Y, which depends on X, also might not be true, and if we go far enough forward, the entire structure crumbles.

Despite these differences of approaches, I would hope we can come together on the view that Christian doctrine should be Christ-centered. It is just such a doctrine that I will discuss in the latter portion of this book.

I hope this little book will help people come out into the open with their discussions of origins. It will be helpful to understand what the foundation of our belief system is. Whether we come to an agreement or not, simply discussing the scriptural roots, theology, and yes, the science of origins will prove valuable as we learn to live in God's great universe.

THE MAJOR VIEWS AND BASIC ARGUMENTS

Details of the various views on origins can get quite confusing. For a brief outline see *Christian Views on Origins*, a booklet in the FastTracts series from Energion Publications.

YOUNG EARTH OR YOUNG AGE[3] CREATIONISM[4]

While I am pretty well known to be a theistic evolutionist, I have argued that this is just one way of viewing the biblical and

---

3 Many who believe in a young earth object to being called young earth creationists and prefer to be called just "creationists" but I maintain the distinction. "Young age" emphasizes that the entire universe is young, not just this planet and solar system.
4 Amongst many others, major representatives of this view include Dr. Henry Morris and Dr. Gary Parker of the Institute for Creation Research (ICR, http://www.icr.org). Parker and Morris are the authors of *What is Creation Science?*

the scientific evidence on origins. I believe that intelligent people hold all of these different positions. That doesn't mean that I think they are all equally well supported by the evidence, either biblical or scientific. Though I will respond with vigor to the arguments of positions with which I disagree, that doesn't mean I think the people who advance them are stupid.

Here are three key elements of a literal view of scripture. A person who accepts all three elements listed below will almost always also accept a young earth and stick with that position.[5]

1. The Bible is to be taken literally where possible (this is a common conservative position, though not all conservative Biblical scholars adopt it.[6] Gleason Archer, to whom I will refer in my discussion of old earth creationism, is a strong advocate of inerrancy and would be considered a conservative evangelical, yet takes a quite different approach. Biblical inerrancy and biblical literalism are not the same thing.[7]

2. The Bible is the decisive source of knowledge whenever it comments on a topic, including science. There can be some variation in what an individual takes as a "biblical comment on a topic." Not all literalists are *literally* the same.

3. The obvious literary form of the Genesis prehistory is narrative history. (Compare Tim LaHaye's view, in which it is to be taken literally if it can be taken literally. One's ability to take something literally may vary considerably.)

Note that those are my restatements of the issue, and not quoted from any particular source. Accept those three things and you will be (or become as you study the Bible) a young earth creationist.

---

5 Since I'm summarizing, let me also recommend reading the presentation of these elements by Dr. Kurt Wise in *Faith, Form, and Time*, Section 1, pages 3-39. I strongly recommend Dr. Wise's book as the one book to read on young earth creationism–if you're only going to read one, make it this one.

6 But note that acceptance of biblical inerrancy doesn't lead directly to a belief that Genesis 1 & 2 are narrative history. An Old Earth Creationist or Theistic Evolutionist can accept the doctrine of inerrancy, and many do. The key is believing that the text is written as narrative history.

7 A good example of a scholar who holds this position is Tim LaHaye, who in his book *How to Study the Bible for Yourself*, chapter 11, page 159, makes it his first rule of hermeneutics.)

Reject them, and you have many other options, but you are unlikely to accept a 6,000 year old earth created in one literal week.

There are those who believe in young earth creationism who will not accept what I have stated here, and will argue that there is good scientific support for their position. But I believe I am being fair, and that the issue does primarily depend on one's view of the nature of the Bible.

Young earth creationists have criticisms of evolution, but the only thing that ties the position together is the Biblical material. The problem is that they lack a complete scientific model explaining the scientific evidence in the context of a 6,000 year total history of planet earth.

The fundamental point that must be addressed in a debate between young earth/age creationists and any of the other groups is the view of what scripture is. The question is not how powerfully inspired a passage of scripture is, but the way in which it is inspired, and the nature of the content.

Now let's look at the basic evidence, taken literally.

» The Bible says the earth was created in one week.

Taken literally and as narrative history, and if one has a bias in favor of the literal reading, this one is pretty clear. Other creation stories, such as Psalm 104, will be read as more general descriptions and the "history" label will be granted to Genesis 1 & 2. Between Genesis 1 & 2, priority will be given to a reconciliation of the accounts.

The idea, for example, that Genesis 1:1-2:4a and Genesis 2:4b-24 have potential contradictions which must be reconciled starts with an assumption that they are narrating a story in some kind of order, that is, attempting to convey the process of creation over a period of time.

When the New International Version employs the translation "had formed" in Genesis 2:19, breaking up a sequence of narrative verbs in Hebrew, the reason is that they want the order of creation to match. Genesis 1 has the land animals created first on the sixth day, and human beings afterward. Genesis 2 would have this process reversed without this questionable change in tense.

» The Bible provides genealogies in Genesis 5 & 11 that provide complete chronological data.

Again, assuming both literal and narrative history, these genealogies provide a very specific answer to questions about the age of the earth. Those YEC advocates who allow up to 10,000 years rather than sticking with 6,000 years depart slightly from the basic interpretive approach by allowing gaps in genealogies that have each person's age at their first son's birth specified, and the number of years they lived after that, but the person specified might be a grandson or great-grandson.

Yet many people make arguments based on the ages of the patriarchs, such as that Noah would have known Methuselah. If the genealogies have gaps in them, this might well not be the case. Thus there is a strong tendency to argue against gaps if one holds a young age position. Having restricted the age of the earth to 10,000 years, it is a minor step to go on and just hold it to 6000 years.

» The Bible provides a narrative of the flood.

I find it odd that some young earth creationists try to develop their model with the flood and related geology separated. If there was a worldwide flood, if the Genesis narrative describes it essentially as history, then it should become an integral part of the theory. An event such as described in Genesis, especially in the priestly account,[8] would definitely leave a mark on the geological record, and one that could be predicted.

Here, again, one's beliefs regarding the sources is often determinative. Do you assume that the flood is history and therefore interpret all evidence in such a way as to support this belief, or do you believe that geological evidence is reliable and primary in determining the history of the planet, and thus interpret Genesis 6-9 in a way that accords with the scientific data?

There is a continuum here, but most young earth/age creationists put almost all the weight on scripture. Science must be conformed to scripture to be true science. For them, no matter what sort of mass of evidence that archeologists, anthropologists,

---

8 Henry Neufeld, *The Creation and Flood Stories*, Gonzalez, Florida: Energion Publications, 2020, pp. 7-12.

geologists, and others gather showing that there are major problems with this chronology, this clear reading of the Biblical record must be decisive.

I would suggest that in discussing this issue, those who disagree should start with where they stand on the three issues of biblical interpretation I listed earlier. Without an agreement on those issues, agreement on derived issues is unlikely.

OLD EARTH CREATIONISM[9]

Old earth creationists differ from young earth creationists primarily on the age of the earth. There is good reason for this. The physical evidence that the earth is more than 6,000 years old is overwhelming. While there may be debates on speciation and on many details of biological evolution, lines of evidence from many different branches of science converge to demonstrate that the earth is old.

But the change in the age of the earth is not so simple. It has an impact on many other aspects of how the Genesis account is to be read.

It's important to understand the vast difference between the "young" and "old" in terms of the age of the earth. Young earth creationists suggest 6-10 thousand years. Old earth creationists accept the age generally accepted in scientific circles, 4.5 billion years. Taking the most common time frame of 6,000 years, that's a ratio of about 1:750,000. Often young earth creationists point to errors in various dating method as evidence that the earth really could be young, but it is important to note that these errors are generally very small compared to the difference between the two time lines. It would be necessary for the errors to be proportionally much greater to suggest that much of a difference.

The key elements of the old earth creationist view are:

» Each day in Genesis 1 represents an indefinite period of time
» God was active in creation throughout that time
» Though there may be considerable variation, and thus evolution, within groups of creatures, major groups are products of creation

---

9   Some key representatives of this view include Dr. Gleason Archer, and Dr. Hugh Ross (Reasons to Believe, http://www.reasons.org).

» As a corollary to this, physical death does occur before the fall, i.e. creatures created on the fifth and sixth days would die
» Humanity is a special creation of God
» The fall changed humanity's spiritual nature, but was not responsible for introducing physical death into the environment

In my discussion of young earth creationism, I mentioned three points regarding the Bible that are accepted by young earth creationists. If one accepts these three points, one must accept a young earth. Old earth creationists hold a modified view of the first and third of these points. They believe that one must determine whether something in the Bible is to be taken literally starting from a neutral position. Gleason Archer, for example, indicates that it is equally wrong to take something figurative literally as it would be to take something figuratively that was intended literally. In his words, "We grievously err in our interpretation when we interpret figurative language literally; we likewise err when we interpret literal language figuratively."[10]

Archer is, of course, no liberal, and in fact is one of the most prominent advocates of inerrancy. The issue here is not the authority or accuracy of the Bible, but rather what the Bible is actually saying. Thus when young earth creationists criticize old earthers for abandoning the Bible, in fact the problem is that the old earthers have abandoned the young earthers' *view* of the Bible. The debate should be about that approach to reading, as without settling that issue, there can be no resolution.

This difference extends to the third point, in that old earth creationists don't view the Genesis story as narrative history. They do, however, view it as containing and assuming certain historical elements. They will provide explanations for the time taken when Adam names the creatures, and also look at how the earth existed under the conditions described in each of the creation days. In other words, while it is not a simple narrative, the Genesis narrative does describe natural history in figurative language.

---

10 "The Witness of the Bible to its Own Inerrancy," quoted from http://www.biblicalstudies.org.uk/article_witness_archer.html [last accessed 09/23/2019].

Most importantly, old earth creationists generally accept the second point, that when the Bible speaks about science it does have priority. They would simply maintain that the Bible makes less statements, and less precise statements, about science.

While old earth creationists generally believe that physical death occurred prior to the fall, they do see the fall of humanity (Genesis 3) as an incident in historical time. Humanity chose to disobey and as a result was separated from God, and made subject to mortality.

Finally, old earth creationists generally hold that the flood (Genesis 6-9) was a local event, not a global one. With the geological record explained by an old earth, there would be little room in the evidence for a worldwide flood.

VARIANTS ON YOUNG OR OLD EARTH CREATIONISM

RUIN AND RESTORATION

The ruin and restoration theory is less prominent today than in the past, but it still has adherents.

I have previously mentioned this theory in the pamphlet *Christian Views of Origins*[11] and in my review of the book *The Invisible War*[12] by Donald Grey Barnhouse. I discuss some of the translation issues involved in my translation and notes on the creation story.

The ruin and restoration theory holds that the current creation is one of a series. Most advocates would hold that there was one creation, then a destruction, and then recreation, though some allow there may be a number that we don't know about. The key basis for this doctrine is a translation of one word in Genesis 1:2, the Hebrew word which practically all translations render "was." The Hebrew word here is "haytah" which is the perfect (suffix) form of the Hebrew verb "hayah" which means "happen," "become," or "was" in most cases. Advocates of the ruin and restoration theory argue that it should be translated "became" here rather than "was"

---

11 Formerly titled *God the Creator*.
12 https://henryneufeld.com/the-invisible-war-review-and-response/ (last accessed 9/24/2019).

and they point to the huge number of cases in which this verb is translated in that way throughout the Bible. The difficulty with this argument is that it ignores the syntax of the passage. The vast majority of the uses of this verb are also used with a different syntax. If one limits one's study to those uses in which the syntax is similar to what it is in this verse, the statistics look much different.

Advocates of this view also bring Isaiah 45:18 and Jeremiah 4:23-26 as descriptions of the destroyed world. Such interpretations ignore the use of figurative language and also take the passages out of context. Both are part of an existing prophetic oracle with a very specific application at a time that is now past, but easily identifiable. An interpreter would need to establish a strong contextual basis for applying these verses to a different time than is clearly the referrent of the passage of which they are part.

What are the advantages of this view? Basically one can hold that the earth is old, which eliminates some of the clearest difficulties of the young earth view. Like old earth creation and theistic evolution, this view also allows for death prior to the creation story. In fact, it allows pretty complete destruction of life on the planet prior to the current creation. At the same time, advocates can take Genesis 1-3 absolutely literally, as long as the one translation change in Genesis 1:2 is allowed.

As I noted earlier, the origin of sin is moved to a time earlier than the Garden of Eden, though this is not an exclusive view of ruin and restoration theory. Many Christians would hold to a rebellion of heavenly beings in what would be prehistoric time on earth. This view creates some interesting additional issues with understanding sin coming into the world and with sin death, which I will address briefly a little bit later.

The disadvantages include the need to explain the recent date of the flood as determined from the genealogies of Genesis 1 & 11. One either has to assume gaps in these genealogies as do old earth creationists, in which case one may be accused of not constructing the text strictly enough, or one must deal with all of the archeological problems that a late date (24th century BCE) for the flood produces. In addition, the interpretation required for the texts in Isaiah 45:18 and Jeremiah 4:23-26 are very difficult to sustain.

This position is largely held by those who accept dispensationalism as a system of interpretation. It is a minority position, but is nonetheless held by a substantial number of Christians, and should be given consideration.

SPIRITUAL INTERPRETATIONS

There are some interpretations that see this story simply as a spiritual illustration, explaining how God provides the logical basis of reality. God is continually the creator, and we derive ideas of how God acts from these stories of primeval times, just as we do so from stories of early patriarchs, prophets, and apostles.

In some ways you might see a continuum between a literal description of historical events and a spiritual illustration, rather than a binary choice. One can believe in a single creation event and still accept continuing creative activity by God, for example.

PREPARING THE LAND OF PROMISE

This view holds that the earth and humanity are much older, but that Genesis 1 refers specifically to the preparation of the Promised Land. This explanation does not interact as with the geological evidence as much as the ruin and restoration theory, because it does not deal with events more than a few thousand years old. It also can be seen as a spiritual interpretation.

INTELLIGENT DESIGN

Intelligent design is the view that there are elements of the devopment of living creatures on earth that cannot be explained by evolutionary processes, and that these are explained instead by the involvement of a designer.

Most who accept some form of intelligent design see this designer as God, though they will emphasize that there is no requirement that what they are observing is divine intervention. It could, for example, be some kind of alien contact or manipulation. Other than that we cannot directly observe this designer, we really can say very little about it as an entity, because science can only observe the process. If it is influenced from outside the only evidence is that there is an appearance of design.

One of the key proponents of this theory is Michael Behe and probably the most famous single book advocating the theory is his book *Darwin's Black Box*. In this he uses a set of biological processes that he believes could not have come about by evolutionary processes and which require some sort of intervention.

Opponents come from two directions. Young age creationists consider Intelligent Design to be simply another variation on evolutionary theory, because it accepts (potentially) an old universe and earth, and because it accepts some of what young age advocates call microevolution. In fact, some intelligent design advocates accept common descent and the role of natural selection.

THEISTIC EVOLUTIONISM OR EVOLUTIONARY CREATIONISM[13]

Previously I've discussed young earth creationism, old earth creationism, and ruin and restoration creationism. That brings us to theistic evolution, or I could say theistic evolutionary creationism.

Though theistic evolutionists may have varying beliefs regard to the nature of God, in general, they see God as the source of all existence in one way or another. Evolution is simply a process which diversifies life in the universe, as much a product of God's activity as any other natural process such as gravity or a chemical reaction. In Christianity, theistic evolutionists can be found in most of the major theological streams. There are people who believe in biblical inerrancy and nonetheless are theistic evolutionists.

Also, there is generally no difference between the scientific understanding of theistic and non-theistic evolutionists. They will generally see very different philosophical frameworks for the events that they study, but the events themselves, and the properly scientific framework for studying them are the same. In terms of science, all three of the other views I have discussed involved some

---

13 Representatives of this view include M. A. Corey, *Back to Darwin: The Scientific Case for Deistic Evolution and* Howard J. Van Till, author of *The Fourth Day*. As I have noted, some, myself included, object to the terms "creationist" and "evolutionist" since neither is a complete philosophical or religious system. The terms are convenient, however.

debate over what may be regarded as natural processes, and some expectation of an identifiable intervention by God in the natural world. While a theistic evolutionist can believe that God can intervene (I do, for example), in general he or she will not regard such intervention as a proper subject for scientific study, because it will not be repeatable. If God performs a miracle, we will see results, but if the cause is outside of the natural realm, that cause would be beyond scientific study.[14]

To the Biblical literalist, there is nothing about theistic evolution that would commend itself. It is not compatible with a literal reading of the first 11 chapters of Genesis. This is one area of debate that can become unnecessarily heated. When a literalist tells a non-literalists that they are "abandoning the Bible" in accepting evolution, what the literalist really means is that the evolutionist is abandoning a *literal reading* of the Bible. For many Biblical literalists, the literal reading is the only possible one, and thus the two are equivalent, but it is important to note that for many, many biblical scholars, there is no such bias.

Old earth creationists read Genesis less literally than do young earth creationists. In particular, they interpret the days of Genesis 1 as symbolic of substantially longer time periods, and take the descriptions of the individual days as much more general looks at what happened over that period of time. While this approach does not take the passages literally, it does take them as historical in some sense. The old earth creationist does not take the genealogies of Genesis 5 & 11 as complete literal history, but they do take the individuals as historical people, and simply assume that there are significant gaps in the lists. Some would also take the ages as symbolic.

For the theistic evolutionist, Genesis 1-11 is not to be taken literally at all. There may be historical events behind some of the stories, but the purpose of those chapters is not to convey literal, narrative history. What they do is present God's activity and his relationship to the universe in terms that would have been comprehensible to the people who first heard and then read them. It may be possible that people described in the genealogies were historical

---

14 In the next chapter I will look a bit at process theology and its potential view of miracles, which blurs some of these lines.

people, but that is not the primary question. The line of connections drawn between the first human being and Abraham, and then from Abraham to the chosen people is the key factor, irrespective of historical details.

This understanding is anathema to Biblical literalists, and makes many Biblical moderates uncomfortable, but it is really an application of a very sound Biblical principle: *Take what is intended literally as literal and what is intended figuratively as figurative.*[15] In this case, one needs to look at the principles, i.e. the message that was encapsulated in these stories that goes beyond the common background material. If one studies the cosmology of the ancient near east and the literature written about it, one will find that it is very compatible with the language of the Bible. The stories and the events are substantially different, because the Bible is slowly introducing monotheism (well developed in the Priestly source of Genesis 1), and the one God it teaches is very different from the pagan gods. But the Bible does not try to change the basic idea of the earth that is round like a dinner plate, floating on the sea beneath with the vault of heaven stretched above it. (See my booklet *The Creation and Flood Stories*[16] and my article "Psalm 104: God, Creator and Sustainer.")[17]

Understanding the part of the message that is timeless is really quite simple. Remove the common elements, and what is newly introduced is the important part, or more precisely they constitute the message that God intends to convey. This is why one can truly

---

15 "Literal" and "figurative" are used here as general labels for a more complex reality. A passage can be figurative in many different ways, and the same passage can have both literal and figurative elements, such as a story that is told as an illustration. The story may largely consist of historical events, but be told in such a way as to reflect the narrator's point of view, and to point to something beyond its historical referents. The discussion really needs to be extended to discussing precisely how you *do* take the passage, which I will address in the next chapter.

16 Henry Neufeld, *The Creation and Flood Stories*, Gonzalez, Florida: Energion Publications, 2020. This book is just 20 pages, 8.5" x 11" and is intended to help readers understand source and redaction criticism, two of the historical staples of biblical criticism.

17 https://henryneufeld.com/psalm-104-god-creator-and-sustainer/ (last accessed 9/24/2019).

believe in biblical inerrancy, and yet accept this figurative view, because what must be understood to be inerrant is the message that the Bible intends to convey, and also allows that the message can be presented through the cultural background of those who write.[18]

Christian theistic evolutionists (and theistic evolutionists are by no means all Christians) do not seek scientific knowledge in the Bible. They will seek God's message and an understanding of how God works with people and interacts with the created universe in a spiritual sense. They expect that this message will be conveyed in terms comprehensible to the culture at the time.

I personally believe that the story of creation might be conveyed in quite a different way in the future as we learn more about science. Will our current explanations continue to make sense? I have no idea how much we have right, in the sense that we have truly comprehended the reality behind what we see, but I suspect we will find new vistas opening up as we learn.

This rather extreme difference in the way the camps understand the scriptures is one of the elements that makes creation-evolution debates so very heated. To a convinced young earth creationist, even an old earth creationist has stepped outside of the scriptural foundations of the faith. It is not just a matter of disagreement on a minor point of doctrine. It is a fundamental difference in the foundation of the faith.

RELIGION NEUTRAL ECOLUTIONARY THEORY

There are some people who accept the theory of evolution and believe that it does specifically exclude God. There are also those who simply don't believe in God and accept evolution. These two positions are different and it is important to distinguish them. Writers such as Richard Dawkins believe that the advance of scientific knowledge has made it impossible for a rational person to believe in God.

Many scientists, while not believing in God, don't believe that science can answer that question. Stephen Jay Gould, an agnostic, used the phrase "non-overlapping magisteria" or NOMA. What this means is that the natural sciences and theology talk about

_____

18 See "A Short Story" on page 43.

different things and thus cannot produce valid conclusions in one another's sphere. This doesn't mean that science and theology cannot interact. Rather, it means that science cannot answer theology's "why" questions, such as why we exist at all, while theology cannot answer the "what" questions of science. Thus, to Gould, the very notion of trying to determine the history of life on earth, or the age of the universe, from the Bible was pointless. At the same time using palentology as the complete source of answers to philosophical questions was also pointless.

Non-theistic evolutionists generally teach the same processes as theistic evolutionists, but would not refer to God as the source of any order or of life, just as many theistic evolutionists would not mix their teaching of science with their religious views about the source of all order. Each topic would be handled separately and taught in classes on the appropriate subject.

A good presentation of this view can be found in *The Blind Watchmaker* by Richard Dawkins, especially chapters 1-3 and 6. Richard Dawkins illustrates some of the difficulties in discussing origins amongst Christians. Because he agrees with some Christians that evolution is atheistic by nature, he contributes to the accusation that those who do not believe in a young earth cannot possibly be true Christians. His idea here is to suggest that Christianity requires one to accept the nonsense, as he would describe it, of a young earth, and thus to prevent intelligent people from accepting it. This is another reason that Christians must discuss these issues openly.

COMMON ELEMENTS OF CHRISTIAN VIEWS

There is some disagreement about what is essential, but the following broad outlines are quite clear in a biblical view of origins as seen from within an orthdox view of Christian theology.

1) God is the creator of everything (Genesis 1:1-2:4a, Psalm 104:24, Hebrews 1:2, Romans 11:33-36).
2) God creates by simple command and His word is certain (Psalm 33:6-9, John 1:1-3).
3) God put personal care into creation (Genesis 2:4b-25).

4) Human beings were created in God's image (Genesis 1:27), they were good (Genesis 1:31), and later they fell from that state (Genesis 3).
5) God created wisely (Psalm 104:24, Proverbs 8:22-31).
6) God continuously cares for His creation (Psalm 104, Acts 17:26-29).

I might note that right there is where the trouble starts!

One thing that Christians who accept evolution have often neglected to do is explain how our understanding of origins fits with the remainder of Christian doctrine. We may hold differing doctrines, we may hold variations on standard doctrines, or we may have our own understanding of how these elements of our belief fit in. Possibly because these explanations seem so obvious to us, we don't take the time to explain the details.

For example, for many Christians the idea of physical death prior to the fall (Genesis 3) is simply inconceivable. They've never entertained the thought. Old earth creationists, ruin and restoration creationists, and theistic evolutionists all share the belief that there was physical death before the fall, though ruin and restoration creationists believe such death came after the fall of Satan from heaven.

God as an absentee landlord is not consistent with the biblical view of God or of humanity, though some passages in Ecclesiastes may provide a counter-argument. The Bible writers universally consider God to be continually present and active in the world God created. They also do not exempt anything from creation.

At the same time, it's important to note that the large amount of time covered by the biblical texts may give a false impression of continual intervention. I refer here to the appearance that something is outside of the normal order of things. One may believe that God is constantly active, without believing that God regularly intervenes to break the flow. For example, I believe that gravity functions because that is God's will. Thus, no God, no gravity, and so God is continually active. When Jesus allows Peter to walk on the water (Matthew 14:22-33), either gravity or some other natural

17

law must not be functioning as expected. That would be a different type of intervention.[19]

Related to this is the question of God's control of everything. The nature of God's intervention is critical. If God's intervention is from outside, in contravention of natural law, and by an incomprehensible will, then the question becomes why God does intervene in some cases and does not in others.

Christian theologians make considerable effort to absolve God of responsibility for evil. This short book is not the place to look at this in detail, but in some sense evil is separated from God's will. If free will is assumed, God gives a good (free will) which is abused as the origin of evil. One can ask whether that truly absolves God.

In fact, one wonders if God wants to be absolved at all. Consider Isaiah 45:7, presented as a direct word from God: "I form the light and create darkness, I make well being, and create disaster. I, YHWH, do all these things." There has been some considerable debate about the translation of the word I have rendered "disaster." It can also be rendered "evil." But the point actually remains the same. In terms of the structure of the poetry here, God takes responsibility for the entire range of results in God's creation. God is not afraid to take responsibility for God's own work.

So there is considerable variation in understanding of the first point with regard to the creation of evil.

God creates by simple command and His word is certain (Psalm 33:6-9, John 1:1-3)

This passage also reinforces the previous. All of everything was created by God, and God accomplishes his word by simple command. Many take this to mean that God cannot use mechanisms, that creation must occur instantaneously as the result of God's command. This would, however, contradict Genesis 1 & 2 on the

---

19 Or it might not be. There are varied views of how such things might take place. Bruce Epperly (*Angels, Mysteries, and Miracles*, p. 88) writes of miracles in this way: "While miracles are not supernatural invasions from the outside, there are moments in which God's love and power bursts forth, not as an alien force, but as the deepest reality of the universe and our lives."

creation of humanity. In Genesis 1:26-27 God simply creates, as he does everything, by speaking, yet in Genesis 2:7, God forms man from the dust and then breathes life into him. The process differs in two descriptions of the same event.[20] The key issue here is that God is absolutely in command. What exists, exists because God wills it and commands it. He can, as he does, command natural laws, and those continue to accomplish his will. Because God doesn't need to be concerned with divided attention, God can be fully attentive to everything at once. One point here that a Christian evolutionist such as myself must deal with is that God was and is present in every moment of the process of evolution; we see the creator in what is created.

God put personal care into creation (Genesis 2:4b-25). Three major stories of creation tell different stories about God's relationship to his creation. Genesis 1:1-2:4a tells the story of command and power; Genesis 2:4b-25 tells the story of personal involvement, and Psalm 104 tells the story of continuous care. This aspect of creation is easy for all of us to miss. We can get so involved in arguing God's power or God's method that we neglect to actually hear the main point of all these stories—how God relates to us.

In addition to asserting continual care of creation, Psalm 104 asserts continual creative activity. There is really no distinction between the activity of creation at a particular time and the general relationship of the creation to the creator. Thus a robust doctrine of creation will need to take into account creative power in the present, whenever that "present" is.

Human beings were created in God's image (Genesis 1:27), they were good (Genesis 1:31), and later they fell from that state (Genesis 3)

This is a key element of the story for a Christian theistic evolutionist who believes in the atonement, as I do. Traditionally, humanity begins in moral innocence, has the opportunity to be in an obedient relationship to God, and then falls away from that state.

This does not mean that they had to live at any particular technological or intellectual level. The primitive human could be

20 Is this an event or a series of events? That is part of the question.
Vocabularly is very important in discussing creation!

in communion with God while creating the first stone tools. Or earlier.

But an essential story of the Biblical story from the Christian perspective is humanity's need for redemption and the sacrifice of Jesus in providing it.

Yet even so it is possible to see these states not as successive events in time, but rather as logical states. The finite is separated from the infinite. Separated from the infinite, the finite will inevitably fail. Atonement brings the divine life poured out in the creation of the physical universe back into oneness with divinity. This may sound strange to Christian ears, but it is important to realize the limitations of our own understanding both of the source of our life and of our destiny. It has not entered into the human mind what God has prepared (1 Corinthians 2:9).[21]

God created wisely (Psalm 104:24, Proverbs 8:22-31)

The fourth creation story (Proverbs 8:22-31) connects to the third (Psalm 104) in claiming that God's creation is wise. What this means is very interesting, but I think at a minimum it means that we can derive valuable information about God from what he has created, how he creates, and how he continues to create. God reveals himself in action.

This again contributes to the difficulty of resolving the problem of evil. If God created wisely, why are there so many not nice things? This should drive us to a deeper study of creation. It is not just the troubles of our own lives, but genocidal attrocities, mass extinctions,[22] and galaxies passing through one another tearing themselves apart. How does this relate to the goodness or the loving care of God?

---

21 There are those who argue that verse 10 contradicts my reading. I believe, however, that the things revealed us by the Spirit are still well beyond our full comprehension. Anything else is arrogance, something not in short supply in origins debates.
22 Note that many scientists believe we are currently in the early stages of the next mass extinction event. Perhaps we should do some thinking about how such events relate to our lives and our theological reflections.

I will hardly solve the problem of evil in the pages of this book, But I want to direct our attention strongly toward looking at all of God's creation. Scripture texts can give us pointers, but we also need to see the reality of what has happened, and what may happen, if we are to have a healthy spiritual life.

God continuously cares for His creation (Psalm 104, Acts 17:26-29)

Again, God didn't start the machine and leave it running. The laws we observe are God's will made manifest. That divine will is so consistent that it (the natural world) can be studied scientifically. Methodological naturalism is simply a stance on studying things in the best way available for their category of information.

We would not have a concept of miracles if there were not natural laws that contrast with the event or action claimed by the miracle story. The sun does not normally visibly alter its movement through the sky, so the sun standing still would be a major disruption. Assuming that such a miracle was performed by altering the rotation of the earth, rather than by an incredibly massive optical illusion, inertia would result in rather massive destruction as objects flew into space.

The continuous care of creation also ties closely to continuous creation. God is not merely the one who created. God is the one who creates. This is a key element of the creation story in Psalm 104, which I have discussed more extensively elsewhere.[23] I like to use a trio of creation stories: Genesis 1:1-2:4a, which expresses power and certainty, Genesis 2:4bff, which expresses personal care and attention, and Psalm 104, which teaches continuous creation and presence.

There are many other presentations of creation in scripture, yet these three express those diverse elements effectively. Proverbs 8:22-36 expresses the connection between this continuous creation and our knowledge and decision making. God creates with wisdom or via wisdom. Wisdom remains a fact of creation. I would further note that wisdom is in this way made a part of creation, something

23 https://henryneufeld.com/psalm-104-god-creator-and-sustainer/ (last accessed 9/24/2019)

we can learn from it, which will be important as we look for a more robust doctrine of creation.

INTERLUDE: WHAT WAS IT LIKE?

When God said, "Let there be light!"
What was it like?

An explosion of sound
Like rolling thunder
Clashing cymbals
Booming drums
Or a wildly cheering crowd?

Or maybe it was glorious music
An engaging ballad,
An organ performance
A symphony
A marching band
Perhaps an explosion of rock and roll.

Perhaps it was a sweet solo,
A Capella words with power
A soprano reaching star high notes
A bass rattling the foundations
A rich contralto
Or a rapper's energy and rhythm.

Or maybe the Word had no sound
An explosion of light and color
Beauty illumined by soundless word
Dreams of mysterious symbols
Sculptures of thought and design
Even substantial structures of emotions.

Even that might be insufficient, so
A blueprint stretching infinitely
Connections intricate and planned

Mechanisms carrying unresisted power
Measurements of incomprehensible precision
A song, a picture, a word, an action, divine.

Or just God's Word.
"And there was."

TOWARD A ROBUST VIEW OF ORIGINS

So what goes into a robust theology of creation? I would suggest the following elements:

1) An expanded and comprehensive understanding of **revelation** and how we learn about divine things.
2) A **hermeneutic** that can be used consistently.
3) An understanding of divine **involvement** the universe that doesn't run roughshod over that universe, at least as a matter of course.
4) A view of creation that integrates reasonably with our theology. In the case of Christianity, a theology of creation that can be **Christ-centered**, i.e. centered in the person of Jesus.
5) An understanding that leads a consistent and livable **ethics**.

In all of these, I hope for a recognition of the limitations of our understanding so that we don't claim to know things that we cannot possibly know.

REVELATION

As with almost any issue, differences often trace back to our epistemology. When the issue is theological, that epistemology traces back to our view of revelation. Where do we acquire what type of information?

For some Christians, denying that the Bible is the primary source of knowledge is to show disrespect to the Bible. This unfortunately leads to giving a priority to biblical statements regarding science and the natural world, even when those statements do not support such an interpretation.

For many young age creationists, the earth must be young because the Bible says it is. But one can only make the claim that the Bible actually says this by making a number of underlying assumptions. One must assume that the Bible writers intended to provide information on the age of the earth (not to mention the universe), or if this information is incidental, that they had access to some sort of accurate chronological data.

The argument then made is that the source of the information provided is God. Setting aside discussions of the nature of inspiration, we at least know that God did not provide all knowledge to the writers. If information is incidental, why should our assumption be that such incidental knowledge would be provided.

Analyzing chronology from poetry, liturgy, and myth and then complaining of the difficulty of the task is not respectful of the text. It is as though one takes up a physics text in order to write a scientific review, and then comments that the language is not poetic. This is not a problem with the text, but one with the reader or reviewer.

But that is only one side of the issue. The division between general and special revelation is deeply problematic. This division tends to be taught in a way that makes special revelation *more* special than general revelation. But the issue is not that God gave us an untrustworthy general revelation and then fixed things up with special revelation. Rather, these two revelations talk speak not only in different ways, but about different subjects.

One might perhaps use the analogy of a physicist who is also an accomplished poet.[24] In physics texts, this scientist speaks clearly and in great detail about the nature of matter, of particles and various forces. This will happen in vocabulary that not everyone can understand.

In poetry, the same physicist expresses a wonderful beauty which she has seen through the complexities of scientific research and expression. People without the training in physics can get an idea of the complexity, simplicity, and beauty of the universe through this poetry.

---

24 Carl Sagan in "Cosmos" expressed scientific concepts in some very poetic language, for example.

It would not be respectful, but rather rude, to attempt to derive the laws of physics from the poetry and then complain of the difficulty and the lack of scientific precision.

A robust view of revelation asks what is revealed and how, and then reads and responds to this revelation in accordance with the best possible answers to those questions. When the rocks, well suited in their placement to tell us about history and chronology, tell us one story, we would do well to listen to that information, rather than trying to replace it with a more literary, poetic, liturgical, and theological explanation which knows nothing of geology.

HERMENEUTICS

I regard the scientific issue of whether evolution has occurred to be a settled one. That is, it is settled scientifically. No matter how settled it may be, however, people can disagree. I do not think the issues of origins are all to be settled in the church. In Christianity there is a valid debate, though I would prefer that our debate centered on the theology and not on the science. Very few in the church can argue science competently or effectively.

As I observed in the previous section, much of this discussion will have to do with hermeneutics, i.e. interpretation. People often try to avoid interpretation, because they think somehow they will get behind all the veils to see the reality.

I have news for all such intellectual pilgrims: It ain't going to happen!

When I went off to college I studied biblical languages because I wanted to get closer to the original. I didn't want to have other people standing between me and THE TRUTH. I hoped that I could get to the place where I made my own judgments and was not dependent on accepting the work of others. But as I learned more, I found that there is always something else that one depends on. To understand the text, one needs to know what the text is. That not only requires dependence on other experts, but it also remains fuzzy—one cannot be absolutely certain of the text, and in some cases one really has no idea at all.

That is the first horseman of disillusionment with one's self-sufficiency. There follows after him lexicography, archeology,

25

paleography, comparative linguistics, literary criticism, and even a bit of accounting. The horsemen of disillusionment are myriad!

You are always interpreting and are connected with and dependent on other interpreters. This is true whether you deal with scripture or the natural world through science. I recall clearly when my father, a physician, but also a radio hobbyist explained to me not just that the way we understood electricity was a theory, but the way in which that theory was formed and how it predicted what occurred. I dealt with interpretation all the time. "But how can I be sure?" I asked. "You can't," he replied. "You interpret the data as best as you can. This interpretation works."

The interpretation of which he spoke has undergone enormous changes over the years, and yet that remains true. This can teach us that we need to realize that we are interpreting, that living by our best interpretation is a viable plan, and that we need to be ready to refine that interpretation.

When we move from strict exegesis of a text such as the Bible and try to form theology, we have a yet more complex situation. Our theology will be impacted by many things, including the findings of science. While these two supposedly separate sources of knowledge deal with different topics in different ways, they both contribute to our understanding of the universe.

To go back to my illustration of the physicist/poet, while I need to read her writings on physics as a science for scientific information, and the poetry for joy, both of these contribute to my understanding of who I am and how I live in the universe.

For those who hate interpretation, let me note that there is, again a layer of interpretation as we interpret the way in which the various interpretations of various topics impact our view of ourselves, of life, and of society.

This why I have always loved Psalm 19, and I also regard it as a unified Psalm even though it is divided into two parts. Those two parts, however, convey a unified central message. God is the creator and this is why he is also the lawgiver. This suggests to me that God speaks in multiple ways, and not just in text.

In the ten commandments, God addresses Israel and starts by God's identity (to the Israelites likely *this particular* god) and the resulting claim on them in particular. This is one reason that it is important to remember that the ten commandments were not initially addressed universally, but rather specifically to the Israelites who were fleeing Egypt. God addresses them and announces that YHWH (God's name) is the God who has freed them. That is the basis for that particular piece of legislation. But while "the" Torah, or the first five books of the Bible, is addressed to Jews, "Torah" (instruction in general) is addressed to everyone.

So why is God the one who can give Torah to the entire world, indeed, the entire universe? Because God is the creator, and creation tells us of him. If we were to address a "ten commandments" to the world, it might begin "I am your God, who created everything, and specifically created you."

This adds another layer of interpretation, that brought by our experience, individual and communal. It is very easy to ignore our own experience and the experience of communities around us as we interpret.

As I write this, there are protests of injustice in the streets of the United States. There is also looting and violence, both by those taking advantage of a bad situation and by the police and government forces. How you react to the way I worded that will say something about your personal and community experience.

I have watched and listened with interest as Christians worship. So many churches are available online. The prayers of white Americans and of African-Americans are somewhat different, even when both would speak against racism. For the white Americans there is a strong tendency to pray against the violence and to call for peace and safety. For the African-American, there is the tendency to call first for justice. This difference arises from our experience. If you were not experiencing peace and stability before, you are unlikely to pray for a return to it.

But experiences again vary widely. As Christians we hear the "brought you out of the land of Egypt, out of the house of bondage" in a very special way. Jesus has brought each and every one of us out of our own spiritual house of bondage, and led us to this place, wherever "this place" may be. So he can address us much as he does the Israelites in combining these two passages: "I'm both your creator and your redeemer."

Note that this is an additional interpretation of an existing passage based on a different experience. Christians do not read the story of the Exodus in the same way as Jews do. African-Americans, brought here as slaves, with freedom hard-won, do not read the Exodus in the same way as white Christians.

Those who focus on science may feel that they are exempt. But science also has its orthodoxies and its interpretations. It is easier to test these scientific positions, and I believe easier to change scientists than believers, but neither group is exempt.

INVOLVEMENT

The question of involvement comes largely on two fronts. On the one hand there are those whose concern is that the idea of evolution is trampling on God's sovereignty. Because there is a mechanism for origins, perhaps even the origin of life itself, we may have less of a tendency to give God the glory which is due. On the other hand, there are those who are concerned that God's involvement is arbitrary and even contrary to God's nature as we expect it from scripture and other sources.

The Illusion of the Uninvolved God

I am honestly astounded by how many people are concerned with infringement of God's sovereignty. It is, in fact, impossible to actually infringe on the sovereignty of the creator and sustainer of the universe. Everything depends on that creator from a theological point of view, and logically if one posits such a creator.

I am not merely convinced that God is involved in evolution, but that God is involved in everything. Thus if evolution works, it works because the Creator of the Universe willed that it would and continues to will that it works in that way. I reject the idea that the discovery of a mechanism for something in nature in any way diminishes God's involvement. So one of the issues regarding creation is precisely *how* God is involved.

28

Some of my more theologically inclined friends may be questioning this one, but God created humanity a little bit less than God (Psalm 8:4), and God allows human beings to make their own choices and plot their own course. God tries to communicate, but doesn't force communication.

God respects us as part of God's creation. This may be a concept that is a bit difficult to grasp. God's effective power is so great and ours is so small, how can respect be possible? I would suggest that God *likes* what God creates ("God saw that it was good" — Genesis 1 passim) and is respecting the result of God's own creative power.

This goes back to the common issue of sovereignty. If God sees my freedom of choice as good, even when I may exercise it badly, and God empowers that freedom of choice, how can this possibly infringe on any sovereignty? That which God desired has taken place! I, the creature, have exercised a power given to me by God.

So what I mean here by respect is that God allows us choices, but God also respects those choices. We often assume that God can do anything, and in His infinity,[25] that is likely close to true. But when operating in finite space and time, God has to meet priorities. So the question is, what is God's highest priority? Is it our safety and comfort? If it is, he should make the world "child-safe" so that we cannot injure ourselves or one another. On the other hand, suppose God valued our intelligence and independent decision making more than our comfort. In that case, God would have to allow our decisions to be independent, to leave us to live with the results of our decisions. Every act taken to make us safer involves a constraint on our decision making or on respecting the consequences of those decisions.

---

25 If we use the term "infinite" of God. I question the value of this claim, even though I suspect it is true, simply because it is not something I can comprehend. "God is infinte," can be translated to "A (more-than)-being that I cannot perceive has an attribute I cannot comprehend." The translation (which I consider accurate) is somewhat less reverent than the source phrase.

My professor for philosophy of religion, Dr. Gerald Winslow[26] told me that I would have to diminish something if I intended to solve the problem of evil. Either God would not be all-powerful, not all-good, or perhaps evil isn't really evil. As a student I denied the requirement, but at least as some see them, I have diminished all three.

The issue here is how we are to define "goodness." To what extent does it involve safety, and to what extent does it allow adventure?

If we are assuming God as creator, whatever is comes as a result of God's action. Action and inaction make no difference here. God cannot really be more or less involved. God is involved in different ways. Thus observing that God allows freedom for things to go wrong, from the freedom of Galaxies to collide to the freedom of human beings to be destructive or to prevent destructive, does not diminish God's sovereignty, and only diminishes our sense of God's sovereignty if the statement is misunderstood.

If you believe in God as creator, you can't get away. I like Tillich's phrase "the ground of all being." If you're "being" you're on that ground. No other option exists.

Miraculous Intervention

On the other hand, there is the question of miracles. Oddly, this one ties into the problem of evil from a different angle. We have just brushed up against the question of why God didn't design the universe in a way that we would regard as *better*. Why is there a universe where tyrants exist?

The opposite question, or at least a question coming from the opposite direction is why God doesn't do something about it if God can.

I can illustrate this with something personal. My wife has been known to pray to find a good parking place. More importantly, she has been known to get one regularly. People comment that

26 From personal conversations. I served as his teaching assistant for a quarter as a college senior, and benefitted tremendously from his willingness to dialogue. He is now Director of the Center for Bioethics at the School of Religion at Loma Linda University.

God must like her because God provides place to park. My wife also has severe arthritis, which causes a great deal of pain. Despite many prayers, God has not chosen to remove the arthritis or the resulting pain. From these two points one can illustrate much of the problem.

Does God fail to relieve the pain because God can't or because God doesn't want to do so? If God is unwilling to relieve the pain, why is God so willing to provide parking places?

To make it much broader, some have asked why it is that God is spending time finding parking places for my wife when children are starving in India. Does God have a problem with priorities?

If we believe that God is infinitely powerful, or even very, very powerful beyond our comprehension, then it doesn't seem likely that God has to choose priorities between parking places and starving children. So the question instead becomes why does God choose not to use that great power to resolve these problems.

One can respond either that God does not intervene, that God intervenes selectively for God's own purposes, or that God's power is limited in some way. In all three cases the pain continues and the children starve.

I am suggesting that in a sense God cannot intervene, though I do not personally think this is a lack of potential, but rather a set of priorities. God has chosen freedom and a general gift of creative energy to the universe, and while it would seem good to us for our comfort to be provided for, in God's view, that would be to negate the very purpose for which God created the universe.[27]

In asserting God's complete involvement, I have managed to assert a certain distance between God and our own desire and choices. I think this effectively makes God more sovereign, not less. God is following a divinely ordained plan, not mine.[28]

---

27 This is extensively discussed in the works of Richard Rice and Thomas Jay Oord, among many others, both in process theology and open theism.

28 Bruce Epperly provides an entire chapter on the miraculous, titled "In Search of the Miraculous" in his book *Angels, Mysteries, and Miracles*, p. 85-104. This vision is based on process theology. I expand on it in my essays "The Hand of God," "The Hand of God - Prayer," and "The Hand of God - Miracles" (https://henryneufeld.com/the-hand-of-god/).

Thus to me the fact that God chose natural selection as the guiding force in diversifying life suggests that God puts a high priority on freedom, and that he does not choose to alter reality for our comfort or to protect us from the results of our own choices, or from more or less random factors such as destructive weather or earthquakes.

This adds a division to miracles, as I discuss in my Hand of God essays (see links above). God likes the natural laws by which he manages the universe. We should not expect miracles to alter that reality for our convenience, nor should we expect them to be necessary to alter the processes of nature or the production of life. The key miracle, apart from existence itself, is that God reaches out to communicate with us. I would also expect that such communication would not be forceful; that God would not intervene to directly alter our minds and understanding.

Let me add a note here. In any basic course in the Philosophy of Religion, students are presented with the problem of evil. God is omnipotent, God is good, yet there is evil. If God is good, one would assume that he would want to eliminate evil. If he omnipotent, he should be able to eliminate it. So what's the solution? The professor will tell you that there is no way to deal with the problem without dealing with at least one of the legs of the triad. You can say God is not omnipotent, and so is unable to eliminate evil. You can say "good" means something other than what we commonly mean by it. Finally, we could decide that evil is not really so bad after all. In a sense, I have done all three here. First, I've suggested that God must have an order of priorities when acting in a finite realm; that limits omnipotence. He can't create a world in which the results of creatures' decisions are respected, and yet also make certain that everyone is comfortable. Second, "God is good" does not necessarily mean that God wants every small animal, or even every person to live a comfortable life. Third, by looking at the positive effects of hurricanes (and I've experienced a number of these lately!) I've questioned whether evil is really evil.

In this system the answer to the question of why the holocaust took place is that evil people made evil choices and took evil actions, and that apathetic people made ineffective choices and did

32

not prevent those evil actions. There were either an insufficient number of good people, or they also made choices that did not effectively stop the evil actions. The solution, therefore, is for people to learn to make better choices. If God solves this problem, he will do so by communication, but the choices and the actions will remain with people. Taking the "reaping what you sow" principle seriously means that we can't assume that God will come and solve our problems for us. God is expecting us to take responsible action ourselves.

Thus evolution shows to us a God who allows freedom in his creation. It's not a safe universe, but it is an interesting one.

CHRIST-CENTERED

It is a specifically Christian-oriented point that a view of origins must be Christ-centered, but I write as a Christian, conscious, to the best of my ability, of what that means.

For a Christian, all theology does center in Christ, in the incarnation. We would not be calling ourselves Christians if we did not start with Jesus, the Christ. Now I haven't suddenly abandoned my understanding of interpretation to claim some sort of unity here. We have vastly different understandings of who Jesus was and what it means to be the Christ. But those understandings tend to reverberate through our theology.

In Christian doctrine, Christ is the center of creation, as expressed in John 1:1-3. This again tells us that, as in Proverbs 8, God's Word, can we say God's wisdom, is imparted to and embedded in God's creation. In one sense, God's incarnation is represented as God becomes an ordinary human, son of Mary and Joseph.[29] In another sense, God is incarnate as the infinite becomes embedded in some sense in the physical universe. "That Word became a physical being, and was at home right there with us."[30]

This becomes a problem when we make a doctrine's validity depend on technicalities that do not connect back to the center. The Christ-centeredness of our doctrine of creation does not depend on chronology, yet some of the greatest debates about this

29 Herold Weiss, *Meditations on According to John*. Gonzalez, Florida: Energion Publications, 2014, pp. 16-17.
30 Paraphrased from John 1:14.

33

doctrine center on that point. Must of the scientific debate is irrelevant here, yet it beomes an issue of fellowship.

This is unfortunate. Let me recommend here the work of Edward W. H. Vick in *Creation: The Christian Doctrine*, particularly the introduction, pages 1-14. I send you there not to learn the true doctrine of creation, but rather to look at how this doctrine ties into others.

ETHICS

Science can tell me that a certain blow will cause death. It can tell me what the results of that death will be. It can, in fact, scare me to death by enumerating the results of what I might have thought were minor actions. Science cannot tell me what I should desire, or to what extents I should be willing to go in order to gain what I desire.

It is a critical failure in Christian ethics to look simply at the text of scripture for ethical instruction but to fail to look at the natural world. In turn, it is a critical failure to assume that I can discover by looking at the natural world precisely how I should behave.

Let me illustrate this simply and briefly. As a Christian I believe murder is wrong. Science can help me understand that. It can help me investigate it. It doesn't tell me whether a particular killing was murder, and if not murder, whether it was in some way justified.

But let us say that instead of committing murder by striking a blow or discharging a firearm, I am the manager of a chemical plant. I deal with dangerous chemicals that can do great harm. Does the Bible tell me to what extent I should care about dealing with these chemicals safely and preventing contamination of my neighbors?

Well, it tells me I should care about my neighbors, while science, the study of that creation naturally, tells me the ways in which I might harm those neighbors, ways that the writers of scripture did not comprehend and had no way to comprehend.

The types of revelation we have discussed all impact the ethics that we produce, but our ethics is the result of interpretation, of integration, and of community. Our doctrine of creation should exted to "what then." In this connection let me recommend another book, Dr. David Moffett-Moore's book *Creation in Contemporary Experience*.

The doctrine of creation impacts our everyday life.

AN EXAMPLE: GENESIS 3 AND THE FALL

Let's take a look at how this comes together in understanding a specific chapter of scripture. For background, read my booklet *The Creation and Flood Stories: An Introductory Aid to Understanding Source and Redaction Criticism*, which looks at the sources in Genesis 1-2 and 6-9.

A fundamental question in dealing with Genesis especially is just what type of literature each passage is. A great deal of the way we interpret a passage depends on the type of literature we perceive it to be. Both young and old earth creationists, for example perceive the first 11 chapters of Genesis to be narrative history in some fashion. The debate between their two positions has to do with precisely how one understands certain terms in the narrative. Old earth creationists, for example, will tend to see more distance betweent he symbols and the reality.

I like the illustration used by Derek Kidner in his commentary on Genesis.[31] On page 66 he discusses the differences in terms of history between the historical description of David's sin in 2 Samuel 11, and the prophetic restatement of that in 2 Samuel 12:1-6. I think that distinction is a good one to keep in mind, but one should also be aware that Nathan's parable that narrated David's sin is intended in some way to convey historical facts, though concealing somewhat their real referent (even David doesn't realize who he is condemning), and clarifying the moral issues involved.

I would like to add a third category here—not intended as historical narrative at all. Gerhard von Rad, in his OTL commentary, tries to present these early chapters of Genesis as heavily demythologized, and indeed compared to their ancient near eastern parallels they are. But at the same time there are many mythological elements remaining, and I believe those elements, along with the

---

31 Derek Kidner, *Genesis*, Kidner Classic Commentaries, Downer's Grove, Illinois: IVP Academic, 2019.

function and message of the story, give us ample justification to read these passages as myth, and to accept them as performing the function of myth within early Israelite culture.

What indicators show me that this should not be read as narrative history? Those who have read my earlier discussions of Genesis 1 & 2 will notice that some of the same reasons apply, but chapter 3 is even easier. In fact, I have some difficulty seeing how so many people can read this chapter and actually expect it to convey narrative history. Kidner's comment that the New Testament writers take it as history (*op cit*, 66) misses the point, I think, simply because as a myth it is well suited to provide the foundation for precisely the type of doctrines Paul especially was presenting. We are separated from God and need to be reconciled. We are separated from eternal life, and must be redeemed by Jesus.

Indeed, one of the most common passages used to read Satan into Genesis 3, and also involved in trying to make it history, is Revelation 12, which itself is pulled out of the narrative sequence. In my study guide to Revelation,[32] I title that section the timeless conflict, because the rebellion of humanity, or in general creaturely rebellion and separation from God and God's saving activity is not limited to a single historical instant.

In this chapter, however, we open with a talking snake. As we will note there is no indication that the snake is anything but a snake, except that he talks. Then we have magic fruit. Notice that the effect comes automatically. At the end of the chapter God has to block the way to the tree of life because if human beings get back there they will obtain eternal life magically. There is mythology removed here, but this is not entirely demythologized!

So in my view the chapter expresses a state, an ontological reality, without providing us a narrative of the process. One could understand this as indicating an instant in which humanity was offered close communion with God and preferred instead to live independently. It could, as Tillich might express it, simply state the separation of the finite being from the infinite ground of all being. In either case the end-state is the reality with which we live, and the reality from which we look to be redeemed. At the same

32 Henry Neufeld, *Revelation: A Participatory Study Guide*, Gonzalez, Florida: Energion Publications, 2005.

time, I think there is a clear sense of something gained as well. Humanity accepted cognition, choice, and moral responsibility. As a result, redeemed humanity will be, I think, greater than a humanity that never went through that experience and never experienced the choice to do right or wrong.

I note here that in looking at evolutionary theory, the challenge to the notion of historicity is underlined for us. The basic indications of a text with symbolic and mythical elements are already there, but the problem is presented to us more clearly when we understand that creation has taken some time. We know that death has taken place historically before there were humans. That drives us to a deeper study.

The combination in this case of a study of God's word expressed in God's created works, put together with the written word as collected over time and accepted by the community, drives us to a deeper understanding of God's word.

I can now imagine that the destiny God has for me is not a restoration of a primitive state. With all these factors worked in, there is no suggestion that primitive is better. Rather, redemption goes beyond—is better than—a restoration of the former state. We have the suggestion that the new Eden may represent something much greater than the first Eden.

SOURCES

This passage has a single source, the J source,[33] and ties closely with Genesis 2:4-25. If you were reading the priestly source alone you would go from Genesis 1:1-2:3, and then go straight to chapter 5, following which you would read about the flood as the first sign that things went bad. In this case, we have the story of the fall, then Cain and Abel, then the crash represented by chapter 6.

A critical result of this, often ignored, is that those for whom the Priestly source was scripture, exclusive of J and E, there was no option of a fall. While we may dispute whether Genesis 3 as such

---

33 J stands for Jahvist, as these terms originate in German. In English, this is more commonly referred to as Yahwist. For more on these terms see my FastTract book *What is Biblical Criticism?* and my previously referenced book *The Creation and Flood Stories: An Introductory Aid to Understanding Source and Redaction Criticism.*

provides justification for a fall that occurs in time, without Genesis 3 there is no textual excuse to imagine it in a temporal sense.

But this chapter is a unity. If there are any borrowing or other sources, they are at the phrase level.

Translation and Notes

Note: This is a personal and perhaps eccentric translation. I always encourage readers to compare any individual translation work to a standard Bible version, such as the New Revised Standard Version.

> ¹Now the snake was more crafty than any of the wild creatures that YHWH God had made, and he said to the woman, "Has God said that you may not eat from every tree in the garden?"

Note several things about the snake. He is not a special creation. He's one of the creatures of the field. Other than being more crafty and able to talk, we get no introduction. I would simply suggest here that when you have talking snakes, you're probably dealing with something other than narrative history.

> ²And the woman answered the snake, "We may eat the fruit of the trees in the garden. ³But regarding the fruit of the tree that is in the middle of the garden, God has said, 'Don't eat from it or touch it, lest you die.'"

It's interesting that the woman immediately moves to put an extra buffer around God's command. If you don't touch it, you can't eat it. Let's be safe. But moral choices will often require us to operate at the limits of moral decision making. For example, as one makes a decision about the morality of stem cell research, how does one operate with a hedge. You have sanctity of life issues on both sides of the equation. You have to make a decision, and you don't get to hedge it very much. Will you eliminate research that could save lives, or will you protect embryos?

Eve wanted a hedge, as many of us do. It is not necessarily more ethical to be more compliant. This is where science has a massive impact on ethics. Science can tell us, in the most neutral

38

way of which we are capable, what the results of our ethicsl decisions may be.

Eve distanced herself from the problem.

> ⁴And the snake said to the woman, "You will certainly not die. ⁵Indeed, God knows that on the day that you eat from it, your eyes will be opened, and you will become like divine beings, understanding both good and evil."

The odd thing here is that the snake turns out to be right, as the story goes on to show. We often try to ignore this, or interpret around it in Christian understandings of this chapter. "Well, they started to die," we say. I would suggest that there is no way out of this dilemma within Christian theology, assuming that we treat this as an event, except an understanding of grace. God intended them to die, but preserved their life instead. God can repent (Genesis 6:6). I think we have the first instance of it here, and I think we're supposed to notice.

Theologically both of these options present issues. If God changes God's mind, we are speaking of the God of open or process theology. If we eliminate the temporality, we have a need to understand again and in a different way traditional redemption theologies.

At the same time note that God had never denied what the snake promised. God simply said, "Don't eat." The possibility is left open that they would become like divine beings, and yet die as a result.

I use the translation "divine beings" rather than "gods" because I think that fits better with the trend of the Torah as we have it now. It was not that they would become gods in the sense of being worthy of worship, but rather than they would share in an aspect of divinity, namely the ability to bring forth either good or evil.

> ⁶When the woman saw that the tree's fruit was good to eat, and pleasing to look at, and desireable so as to gain wisdom, she took from its fruit and ate it, and she also gave it to her husband with her, and he ate. ⁷And the eyes of both of them were opened, and they knew that they were naked, and they stitched together fig leaves and made themselves loin cloths.

The woman "saw" that the fruit was good. We have an abbreviated narrative. Somehow the snake makes the woman see the fruit in the way he wants her to see it. This passage makes me wonder if we don't have more of a narrative of internal struggle, the sort of struggle that takes place in any child who is contemplating something forbidden. It might be the cookie jar. Indeed, the cookies will taste good, and the child will experience pleasure from eating them, but there is a reason not to. An internal conversation convinces the woman that this is a pleasure worth having.

Conversely, the text doesn't tell us that the woman decided that God was wrong, even though that is what the snake had told her. She convinces herself that the fruit is good, and God's statements about it recede conveniently into the background.

> [8]Then they heard the sound of YHWH God walking in the garden in the cool time of the day, and the man and his wife hid themselves from YHWH God among the trees of the garden.

The immediate result of stepping out on their own is that the human couple are afraid. Notice that God is merely going for a walk, presumably looking to talk with the people he made and placed in the garden. He's not blustering, throwing thunderbolts, threatening, stomping, or anything similar. He's just taking a walk. Humanity has stepped out independently, but is afraid of the results.

> [9]And YHWH God called out to the man, "Where are you?"
> [10]And the man said, "I heard you in the garden, and I was afraid because I was naked, and I hid."
> [11]And God said, "Who told you that you were naked? Have you eaten from the tree that I commanded you not to eat from?"

The human couple had been naked since they were created, but suddenly it becomes important. With self awareness comes shame, shyness, uncertainty of how to present oneself. It's something they will have to deal with on this new path they have embarked on.

> [12]Then the man said, "The woman whom you appointed to be with me, she gave me fruit from the tree, and I ate."

[13]So YHWH God said to the woman, "What is this that you have done?"

But the woman said, "The snake led me astray, and I ate."

Who says the Bible isn't relevant? This scene takes place in myriads of households, myriads of schools, and myriads of workplaces every day! We're confronted by something that has gone wrong, and everybody looks for the person who is to blame. Everyone points at someone else. It can't possibly be our own fault.

Notice that God doesn't ask the snake anything. Is it possible that the snake is simply a symbol for an internal struggle, that God doesn't deal with the snake because it's being used by the woman as a "devil made me do it" kind of excuse? I don't know, but I suspect there's a reason why the snake doesn't get to defend himself.

[14]So YHWH God said to the snake, "Because you have done this, you are more cursed than any of the wild creatures. You will crawl on your belly and eat dust as long as you live. [15]And I will place hostility between you and the woman, and between your descendants and hers. Her descendants will bruise your head, but yours will bruise her descendants' heel."

I have no problem in Christian theology reading back into this passage some reference to redemption, as long as we recognize that we are reading back, but that is not the point in its original context. The passage here simply explains why snakes are considered dangerous, looked down on, and crawl on their bellies. They did a bad thing here and they are paying for it! Women have a feud with them. This is hardly the serpent of Revelation 12, cast down from heaven, or the great Leviathan, portrayed as conquered by God in Psalm 74:13-14.

[16]To the woman he said, "I will make childbearing much more difficult for you. You will bear children in pain, yet you will desire your husband, and he will rule over you."

Again, a description of real life in the real world of that time at least. It doesn't mention good pain medications or women's liberation, but the equality of male and female is something promised in Jesus, after all, and not that much a reality in the history of the world thus far.

My wife tells me that if men had to experience the pain of childbirth there would be no humanity, and I pretty much agree with her. Somehow women keep undergoing the torture and propagating the species. Note here how science speaks contrary to a temporal reading and thus impacts our scriptural hermeneutics. There is no scientific evidence of a time when childbearing was painless, and considerable reason to believe there was no such time.

> [17]To the man he said, "Because you listened to your wife's voice, and you ate fruit from the tree about which I commanded you, 'You shall not eat from it,' the ground will be cursed on account of you. You will eat from it only by hardship as long as you live, [18]and it will bring forth thorns and thistles for you, and you will eat vegetables from the field. [19]You will get bread to eat by laboring until you sweat until you return to the ground, because you were taken from it. Dust you are, and you will return to dust."

The man gets to work hard to produce food. But I think there is a spiritual dimension to this in that having given up total dependence on God he becomes dependent on himself. From now on he must make his own moral decisions as well as producing his own food, building his own shelter, and clothing himself and his family. Independence comes at a price.

> [20]So the man called his wife's name Eve, because she was the mother of all people who were alive.
> [21]Then YHWH God made coats of skins for the man and his wife, and he dressed them.

It's interesting that Adam just now notices that Eve is the mother of all living. Perhaps it was of less importance before they were aware of their situation. In any case, to cover their nakedness, and prevent shame now that they were aware of it, they are clothed.

> [22]And YHWH God said, "Look! The human has become like one of us, understanding good and evil, and now, [we need to take action] lest he should take also fruit from the tree of life, and eat it, and live forever."
> [23]So YHWH God sent him out of the Garden of Eden to cultivate the ground from which he had been taken. [24]And

he dispossessed the man and made him live to the east of the Garden of Eden, and he placed Cherubim with flaming swords turning this way and that to guard the way to the tree of life.

This is another "magic fruit" instance. There is a tree which God must prevent the human couple from reaching, otherwise they may become immortal contrary to God's will. Surely this is not intended as narrative history! Symbolically, this says that God does not provide eternal life to those who are operating in complete independence from him, but the fact that the couple do not die, even though God had said they would, shows that he graciously extends life.

A Short Story

Let me close with a humorous short story. I hope it helps bring my point home.

Genesis Wasn't Written That Way

After some weeks on the mountain, the Lord looked over Moses' shoulder.

"In the beginning God created the heavens and the earth."

"What are you doing, Moses?"

"Well, Lord, I'm trying to write up that stuff about how You created everything."

"I think there could be a few problems with what you've written there."

"Problems? It's just the introduction! And You did, didn't You?"

"Well, apart from the fact that readers many centuries from now are going to debate whether you mean 'in the beginning God created' or 'when God began to create'—this whole writing without vowels thing does leave room for ambiguity—it's not really balanced."

"Balanced?"

"Yes, the 'heavens' and the 'earth' are just not quite of similar size, weight, and importance."

"I don't understand."

"I know you don't, Moses. Let me explain." There was a pause. It was not that the Almighty didn't know what he wanted to say, but

43

even His great servant Moses might have trouble understanding. "Your 'earth' is just one tiny world among many. Let's call it a planet. Your 'sun' doesn't 'rise' in the east and set in the west. Rather, your 'earth' rotates on its axis. In turn, it orbits—that means 'goes around in a huge circle—the sun. All those stars in the 'heavens' are just like your sun. Many of them have planets themselves. Some of the planets right here in your solar system—that's all the planets that go around your sun—are bigger than your earth. There are lots and lots of these stars and planets. Your earth is really a very small thing."

"Lots? You mean hundreds?"

"More than that, Moses."

"Thousands?"

"More."

"Tens of thousands?"

"Lots more. I don't think you know any numbers that are big enough."

"Oh."

"You can see that it's a little silly to refer to 'everything' as 'the heavens' and 'the earth', can't you?"

There was a pause again. "So what do You want me to do Lord?"

"I want you to explain it to them. You'll tell them about all these wonderful things, and then you'll tell them that I created all that!"

Moses muttered something.

"What was that, Moses? I didn't hear you!"

"Pardon the disrespect, Lord, but I was saying that we'd be lucky if they listened for that long."

"But you'll make them understand! You'll explain it and they'll listen!"

"But I don't understand it myself. It's obvious to me that the sun rises in the east and sets in the west and passes under the earth to return to its circuit. Everybody knows that!"

"So what do you suggest?"

Later the Lord looked over Moses's shoulder.

"In the beginning God created the heavens and the earth. And the earth was formless and empty, and God's wind was moving over the face of the waters. And God said, 'Let there be light!' ..."

"That's not how I did it," the Lord muttered. "But I guess Moses knows his business!"

RESOURCES

Achtemeier, Paul J. ed. *HarperCollins Bible Dictionary*. San Francisco: Harper Collins Publishers, 1996. ISBN: 0060600373. See especially the article "Sources of the Pentateuch."

Alexander, David ed. *Eerdman's Handbook to the Bible*. Grand Rapids: William B. Eerdmans Publishing Company, 1973. ISBN: 0802806392. See the notes on Genesis 6-9 and the introduction to the Pentateuch, pp. 122-126.

Metzger, Bruce M. ed. *The New Oxford Annotated Bible with the Apocryphal/Deuterocanonical Books*. New York: Oxford University Press, 1994. ISBN: 0195283562. See in particularly the introduction to the Pentateuch (XXXV and XXXVI) and the notes on the flood story in Genesis 6-9.

Lentz, Geoffrey and Neufeld, Henry. *Learning and Living Scripture: A Guide to the Participatory Bible Study Method*. Gonzalez, Florida: Energion Publications, 2010.

Neufeld, Henry E. ed. *What Is Biblical Criticism?* Gonzalez, Florida: Energion Publications, 2020.

_____, ed. *I Want to Study the Bible*. FastTracts. Gonzalez, Florida: Energion Publications, 2020.

_____, ed. *Christian Views on Origins*. FastTracts. Gonzalez, Florida: Energion Publications, 2020.

_____. *When People Speak for God*. Gonzalez, Florida: Energion Publications, 2007.

Noth, Martin. *A History of Pentateuchal Traditions*. Scholars Press, 2000. Originally published in German in 1948.

Perrin, Norman. *What Is Redaction Criticism?* Eugene, Oregon: Wipf and Stock, 2002.

Suggs, M. Jack ed. *The Oxford Study Bible, Revised English Bible with the Apocrypha*. New York: Oxford University Press, 1992. ISBN: 0195290003. See in particularly the introduction to the

Pentateuch (pp. 7-9), the notes on the flood story in Genesis 6-9 and the introductory·article "Literature of the Ancient Near East" pp. 57-67.

Thompson, Alden. *Inspiration: Hard Questions, Honest Answers.* 2nd Revised Edition. Gonzalez, Florida: Energion Publications, 2016.

Vick, Edward W. H. *From Inspiration to Understanding.* Gonzalez, Florida: Energion Publications, 2011.

von Rad, Gerhard. *Genesis, Revised Edition.* Philadelphia: The Westminster Press, 1972. ISBN: 0664209572.

Weiss, Herold. *Creation in Scripture.* Gonzalez, Florida: Energion Publications, 2012.

_____. *Meditations on According to John.* Gonzalez, Florida: Energion Publications, 2014.

ALSO FROM HENRY NEUFELD ON CREATION

In full color, 8.5" x 11", this book helps you see the creation stories as scholars might see them using source and redaction criticism. It's a great companion both to this book and to the FastTracts booklet *What Is Biblical Criticism?* ($9.99 print; $2.99 ebook.)

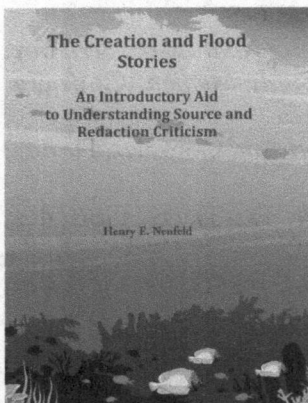

The Creation and Flood Stories

An Introductory Aid to Understanding Source and Redaction Criticism

Henry E. Neufeld

Christian Views on Origins outlines the summary information provided in the first part of this book. It is a valuable resource for churches that want to accept dialog on creation and benefit from people with a variety of views. ($2.99 print; $0.99 ebook.)

CHRISTIAN VIEWS ON ORIGINS